(Photo NIMH)

Cruiser
HNLMS Tromp

The *Tromp*-class was a class of light cruisers of the Royal Netherlands Navy. Originally the ships could not be called "cruiser" for political reasons. They were designed as "flotilla leaders" and their intended role was to be the backbone of a squadron of modern destroyers that was planned at the same time (only one of those was completed before the war broke out). The ships were ordered in 1935; *Tromp* was launched in 1937, and her sister ship *Jacob van Heemskerck* in 1939.

A brand new 'Tromp' during high speed trials in 1938. (Photo NIMH)

7 is the lucky number...

The cruiser *Tromp* was in 1938 the 7th ship of the Royal Netherlands Navy to bear this name. In 1945 an Australian reporter wrote an article on board. He mentions that the crew called her: "The Lucky Ship".

When based in Australia the ship also acquired the nickname: *"The Ghost Ship"*. This was the name given to her by the Aboriginals because the Japanese claimed no less than five times that HNLMS *'Tromp'* had been sunk, whereafter the ship returned to base safe and sound. No single ship, with the exception of the British carrier HMS *'Ark Royal'*, has been claimed to be sunk so often.

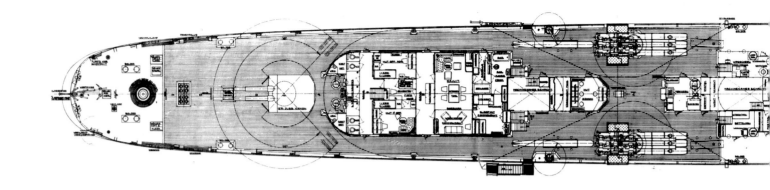

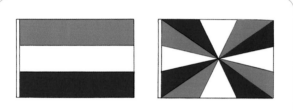

Left: The national ensign, carried by all ships of the Royal Netherlands Navy. In harbour or at anchor it was customary to be worn at the staff on the quarterdeck, but at sea it used to fly from the mainmast.

Right: The jack

The commissioning pennant, worn from the day a ship commissions until she is de-commissioned.

Tromp was build at The Netherlands Shipbuilding Company in Amsterdam. (Photo NIMH)

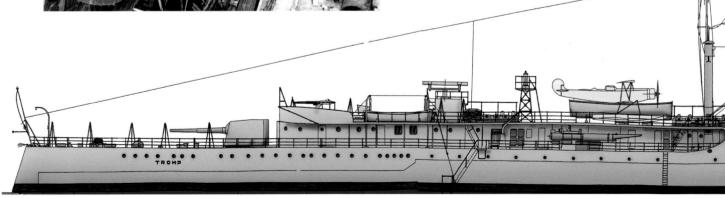

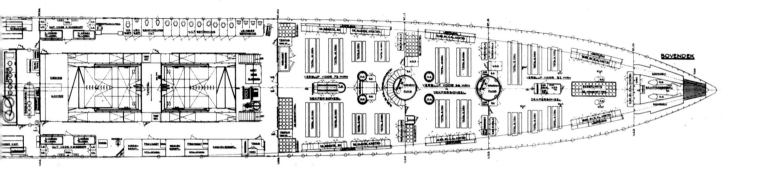

The ship

HNLMS *'Tromp'* was the first flotilla leader of the so-called "Deckers' Fleet Plan", which was passed in 1931. It lasted until 17 January 1936 before the first keel plate was laid down in the yard of the NV Ned. Scheepsbouw Mij. (The Netherlands Shipbuilding Company Limited) in Amsterdam. The launching ceremony was on 24 May 1937 by H.M. Queen Wilhelmina and on 18 August 1938, HNLMS *Tromp* was commissioned by the commanding officer, Captain L.A.C.M. Doorman on 18 August 1938.

It was only during the building of the ship that it was decided to build a second flotilla leader. Building of this warship, which would be named *Jacob van Heemskerck*, would begin as soon as the first ship had been launched. Both flotilla leaders would have a displacement of 3450 tons and a main armament of six 15 cm guns. Mostly, both ships have been considered to be particularly successful. The design was from the engineers G. 't Hooft and W.M. den Hollander assuming that the ships would have to operate continuously in tropical waters. They were renowned for their graceful, elegant lines. Also for their speed and fire power, which was considerable for ships of this dimensions.

Strictly speaking, the new design was superior to a large destroyer. The design incorporated 17 watertight compartments. The ship had light armour on both sides of the hull, on the decks and ammunition hoists, according to a design that had also been used for the larger cruiser HNLMS *De Ruyter*. Hardened steel of the Krupp Works had been imported for protection. Additionally, much aluminium, from Switzerland, had been applied.

While earlier ships were customarily rivetted, a great deal of welding was used during construction.

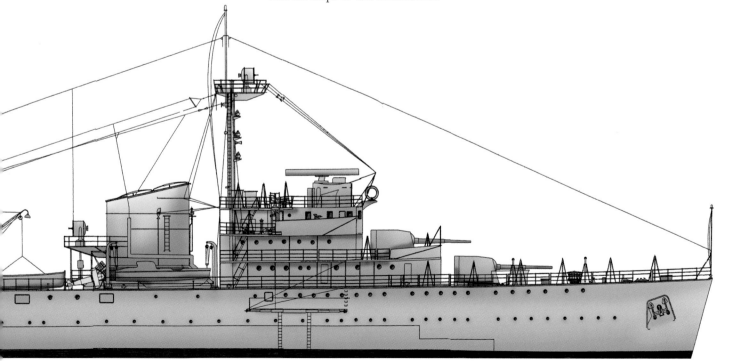

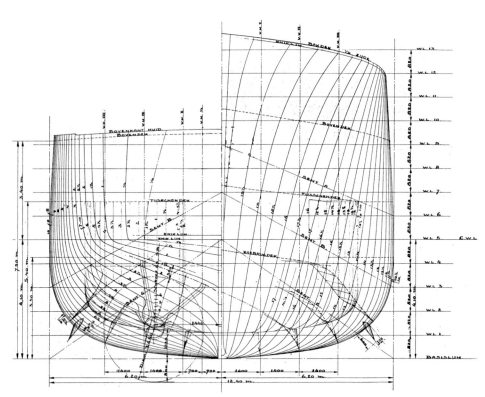

The design of the two ships fulfilled the twofold purposes for carrying out their future duties. In the first place they were intended to strengthen the destroyer fleet in the Netherlands East Indies. At the same time they had to add weight against the ever greater strength of the Japanese torpedo boat (destroyer) fleet.

For that reason the new ships had to be fast enough to keep up with the destroyers. This meant that the armouring of the ship was limited. Even so, 450 tonnes of hardened steel was used

The result was vertical and horizontal armouring which at the thickest parts was 15 mm and 25 mm respectively, providing protection to the most vital parts of the ship. Although limited, because heavier armour would certainly have had a detrimental effect on speed.

Apart from the hull, guns, fire control, communication and ammunition hoists were armoured also.

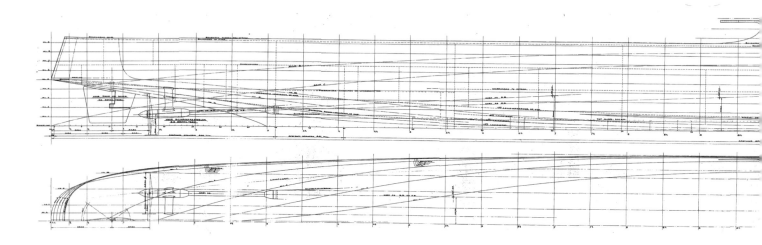

Plans

Modelling plans are available through the Dutch Modellers Association (NVM):

Plan No. 10.11.016
Scale 1/200

www.modelbouwers.nl

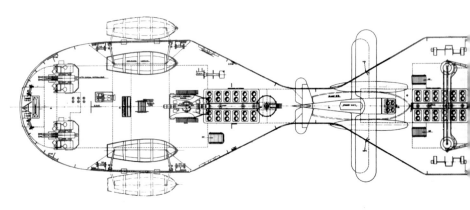

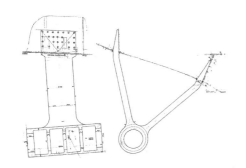

Some details of rudder and props.
The ship was propelled by geared turbines driving twin screws, the total power output being 56.000 s.h.p. for 32 knots maximum speed, though she actually reached 33,5 knots on trials. (Photo NIMH)

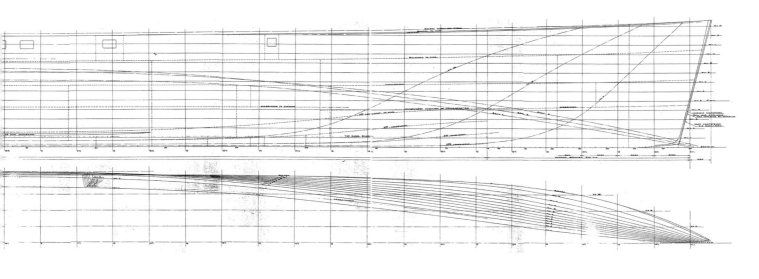

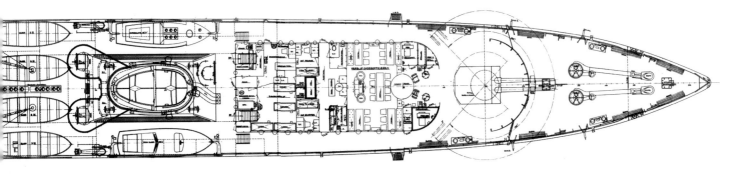

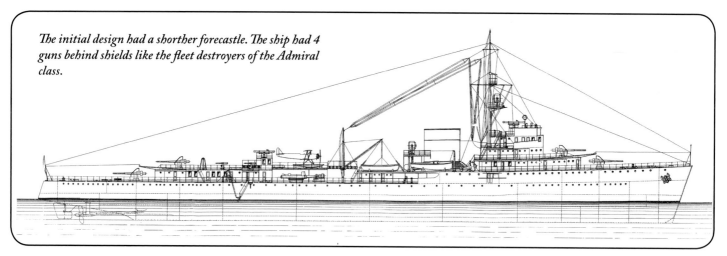

The initial design had a shorther forecastle. The ship had 4 guns behind shields like the fleet destroyers of the Admiral class.

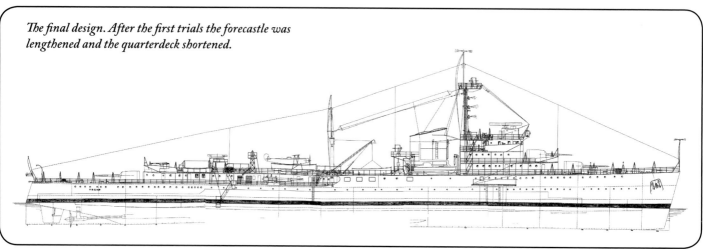

The final design. After the first trials the forecastle was lengthened and the quarterdeck shortened.

After launching, the ship was completed at the quay. *(Photo NIMH)*

Technical data

Displacement:	3,787 tons	(4,225 tons full load)	
Measurements:	length over all	131.95 m.,	(433 ft.)
	beam	12.43 m.,	(40.75 ft.)
	depth	7.5 m.,	(24.60 ft.)
	draught	4.47 m.,	(14.63ft.)
Complement:	295, during WW2 it was 393.		
Machinery:	56.000 hp, 32 knots (33,5 knots on trials)		

The building cost of HNLMS Tromp was Fl. 7.700.000,-- (Euro 3.500.000,--)

During the Second World War HNLMS *Tromp* was classified as a light cruiser. The Royal Netherlands Navy therefore decided in conformity with British light cruisers to appoint a Captain as C.O.

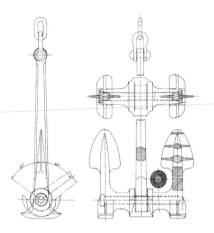

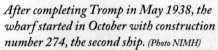

After completing Tromp in May 1938, the wharf started in October with construction number 274, the second ship. (Photo NIMH)

Tromp was built in Amsterdam at The Netherlands Shipbuilding Company. Her launching ceremony was marked by the presence of H.M. The Queen, who christened this new unit of Her Navy.

Tromp shown as she first appeared. Classed originally as a flottilla leader, this category changed to light cruiser, because of displacement, armament and other characteristics she was more suited to the latter classification.

The surface ships were destined for reconnaissance; it was considered that they were sufficiently heavy armed to take the offensive if the opposition met was not too heavy, and sufficiently fast to break off the engagement if they in turn were engaged by stronger enemy forces. Under these circumstances they would have to hit and run, while running attempt to lure their opponents to be attacked by submarines.

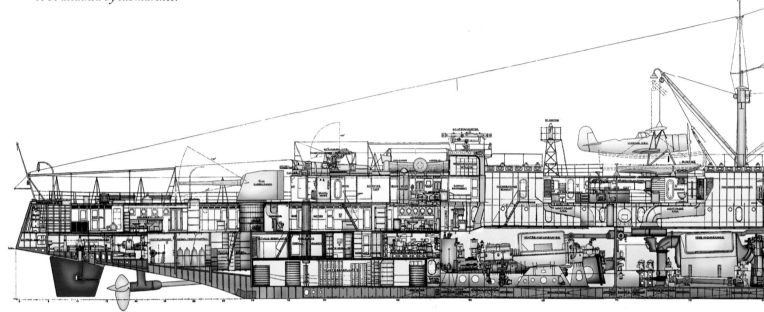

The ship was propelled by geared turbines on twin screws. The actual power output being 56,000 s.h.p. for a speed of 33,5 knots on trials. (Collection Jt. Mulder)

Engines

Propulsion: two Werkspoor Parsons geared turbines of 56.000 HP. Steam was obtained from four 'Werkspoor Yarrow-type' tubular boilers. They turned out to be reliable machinery by which two three bladed propellers of 3.9 meters diameter were driven. The ship had two boiler rooms. The starboard propeller shaft ran to the foreward engine room and the port propeller shaft to the aft engine room. However the foreward engine room was a deficiency in design. There the temperature could become very high, because all boilers were situated close together.

During the trials in Scottish waters an ultimate speed of 35,05 knots was reached. Oil fuel: 860 tons. The radius of action was 5000 miles at a speed of 10 knots and 1400 miles at a speed of 32 knots.

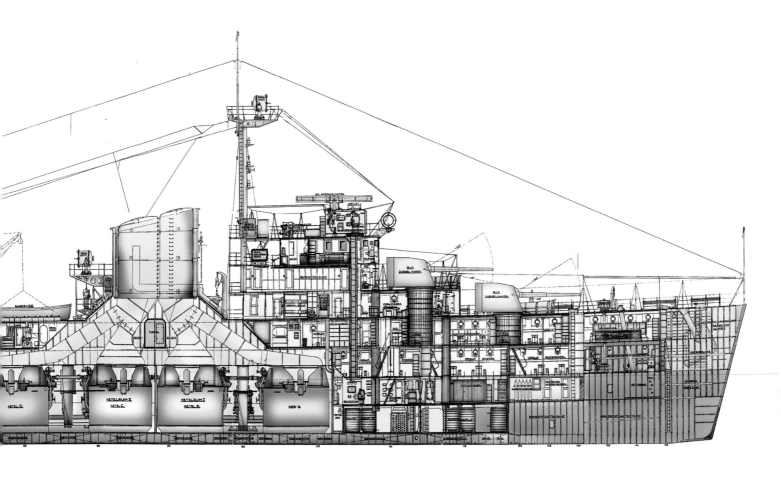

Left:
Work in progress on the large funnel.
(Collection Jt. Mulder)

Above:
Leaving the port of Amsterdam in 1938 for
trials. (Collection Jt. Mulder)

Armament

Considering the relatively small displacement of 3787 tons, she was rather heavily armed with six 15 cm guns in twin mountings. These were fitted behind shields because turrets would have added too much weight. The 15 cm cannon No. 11 could be elevated 60 degrees and had an effectual range of 17.500 meters. They were built by Wilton Fijenoord from a Swedish design under licence of the firm of Bofors. For the ammunition supply each installation had two ammunition elevators.

The aft section, with one of the two Bofors twin mountings. Shortly before the outbreak of the war this construction was modified.
(Photo NIMH)

Aft on the awning deck anti aircraft armament of four 40 mm Bofors machine guns in twin mountings and fire control was installed. The entire arrangement was of Swedish-Dutch manufacture. At the time the best in the world concerning anti-aircraft artillery. Both, the machine gun and the fire control were stabilised on three axes, the Bofors-Hazemeyer system. In addition the ship was given anti-aircraft armament of four 12,7 mm guns on the upper bridge. The ship was also fitted with two mountings of three torpedo tubes. It would appear later that these were situated somewhat too high. Each launching arrangement had a torpedo davit to move the torpedoes with a separate electric torpedo winch. The ship carried 12 torpedoes. Sometimes spare torpedoes were stowed on the upper deck.

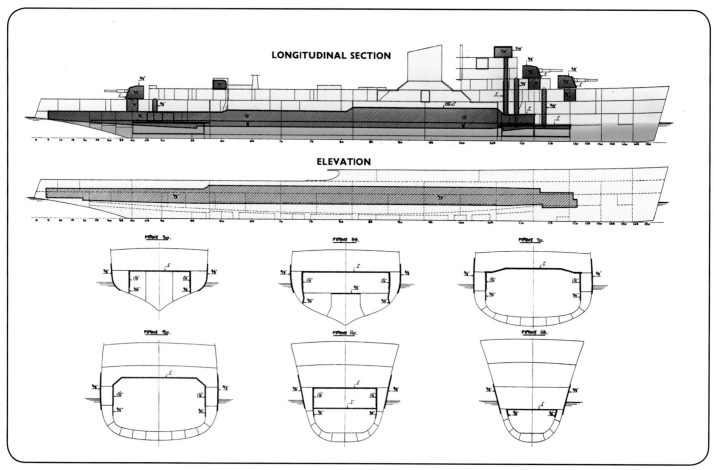

The protection of this light cruiser. No less than 450 tons of hardened armour steel were used in the construction. With a vertical armour of 15 mm (5/8 inch) to 30 mm (1¼ inch) and a horizontal armour of 25 mm (1 inch) to 30 mm (1¼ inch). The armour being extensively used to act not only as protection, but also reinforcing ship's structure.

An interesting view, taken from the platform of the foremast. Aft, just visible the two twin Bofors guns abreast. Shortly before the war broke out in the Pacific, this section was reconstructed in order to get a much larger arc of fire for the aft 15 cm gun. Therefore the two Bofors mountings had to be replaced and were moved to the centre line.

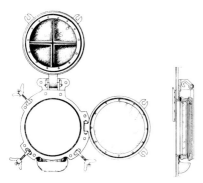

A float plane was to be carried. The ship would not be equipped with a catapult to launch the 'plane. When the 'plane was to be used it would first have to be lowered in the water with derricks. After which it would take off from the water. When the sea plane had landed again, it would be recovered by derrick. For this an electric winch was fitted on the awning deck. When the 'plane was not carried, an extra lifeboat, a B2, could be taken. The derricks and the electric winch were also used to lower and hoist the lifeboats.

Electricity was generated by three turbo-generators. Two diesel generators for general use and an emergency diesel generator on the upper deck. The current generated was 220 volt DC.
A number of internal telephone connections were also installed. There were gunnery 'phones, mooring 'phones, engine room 'phones, etc.
Communication was usually done by radio. Shortwave transmitters and receivers were German made. Much of these were of the Telefunken brand. For encoding messages cryptographic equipment was used based on the German Enigma. The radio call-sign[1] of HNLMS *Tromp* was PAAO.

(Collection Jt. Mulder)

---

1   *International Call-Sign or Signal Letters*

Pendant numbers

During the Second World War, when detached with the Royal Navy the visual call-sign of the ship was: D 28. The Americans used C 39.

After WW2 it was allocated a Dutch call sign: KL 2, which was changed to: C 804 (Nato 15 October 1950).

When HNLMS Tromp was classified immobile, A 878 was allocated (1 April 1955).

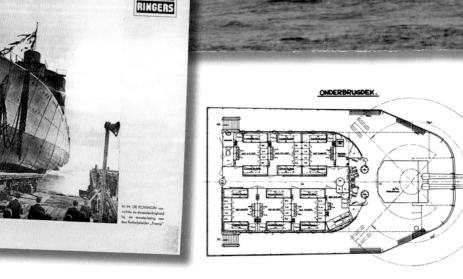

An old magazine of May 1937 reports the launching by H.M. the Queen. (Collection Jt. Mulder)

Even when proceeding with high speeds, the quarter deck keeps dry during trials in 1938.

(Collection Jt. Mulder)

TOP COMM. TOREN.

BRUGDEK.

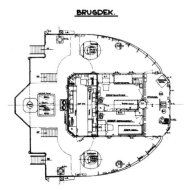

VUURLEIDINGDEK.

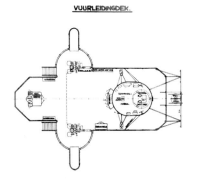

ZOEKLICHTBORDES AAN VOORMAST.

Gun: 15 centimeter Mark 11

The main guns had to be used against surface targets and had a limited role as AA-weapon. But in this last role the mounts proved to be too cumbersome in practice. There were several developments of these guns. The Mark 11 mountings allowed 60 degrees elevation. The shields were 15 mm.

In HNLMS *Tromp* the 'A'-gun was manned by a crew of the Royal Netherlands Marines.

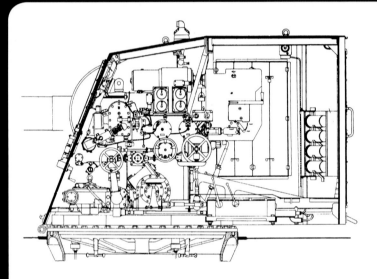

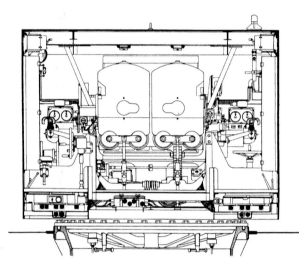

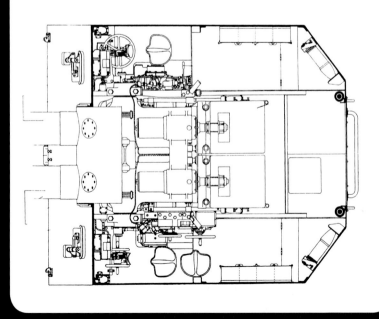

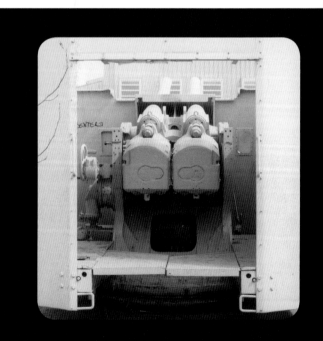

Length	50 calibers
Gunweight	7,5 tons
Initial velocity	2953 feet/sec
Rate of fire	5 - 6 rounds minute
Shell specs	
Shell types	Dutch HE and AP, British HE
Shell weight	Dutch HE: 46,0 kg
	Dutch AP: 46,7 kg
	British HE: 45,3 kg
Range	23,200 yards, elevation 29 degrees
	30000 yards, elevation 45 degrees

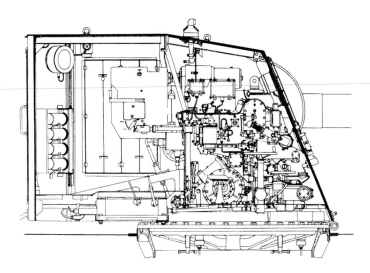

Both photo's show a 15 cm (Mk 8) mounting at the Naval Museum, Den Helder. This gun was removed from the gunboat 'Van Speijk' and is slightly different from guns of Tromp.

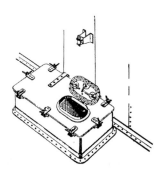

Above:
A visual check of the horizon. Signalman trains telescope.

Left:
The ship berthed in Amsterdam just before commissioning. (Collection Jt. Mulder)

Pre-war photo of HNLMS Tromp in home waters.
Unfortunately this photograph is retouched by the photographer. (Collection Jt. Mulder)

Commissioning of HNLMS Tromp on 18 of August 1939.
(Collection Jt. Mulder)

The wooden bench on the quarterdeck was a gift from the shipyard.

(Photo NIMH)

(Photo NIMH)

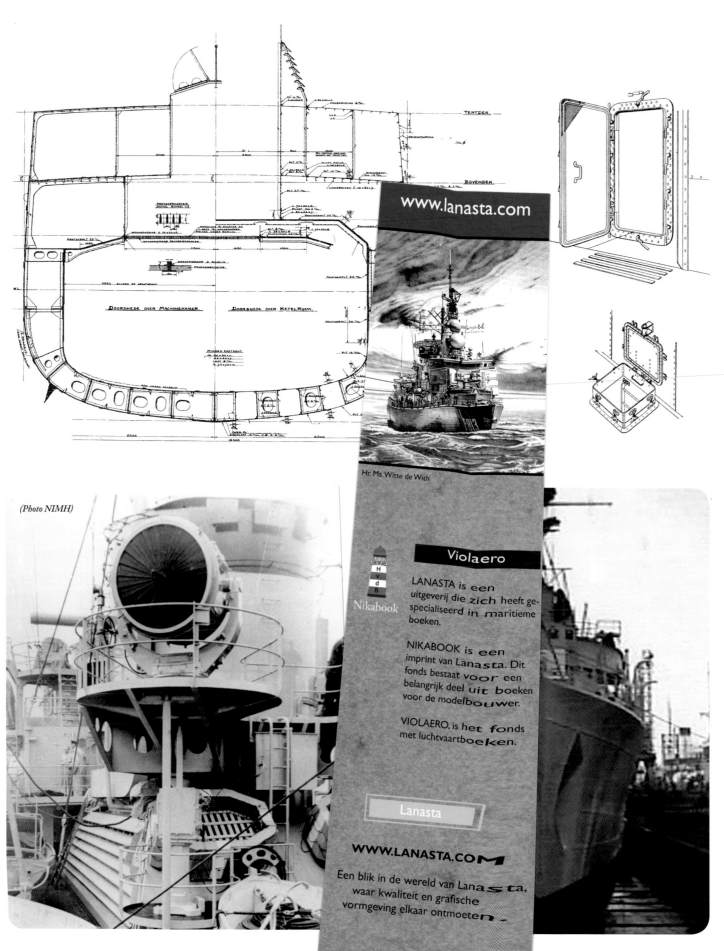

(Photo NIMH)

www.lanasta.com

Hr. Ms. Witte de With

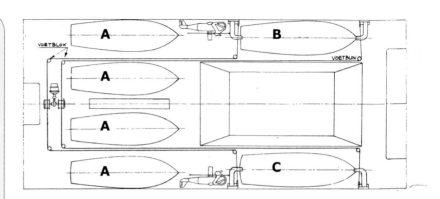

Boats

There where a number of boats on board for transporting stores and personnel around naval bases and anchorages. In the Royal Netherlands Navy these were indicated with a letter, followed by a number. The number stands for the dimensions of the boat.

-**A**- The **B3** boat. Three to four were carried. When no aircraft embarked, a larger **B2** boat instead.

-**B**- Motorboat (1 boat)

-**C**- The boat of the commanding officer. A beautiful constructed motorboat.

-**D**- Two lengthened jollyboats aft.

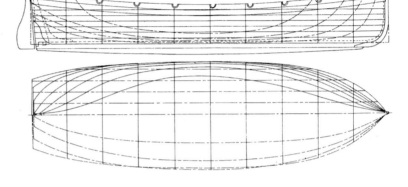

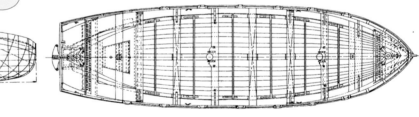

(A) The B2 – B3 boat.

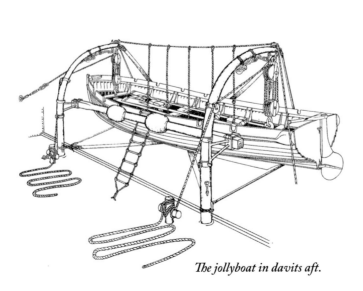

The jollyboat in davits aft.

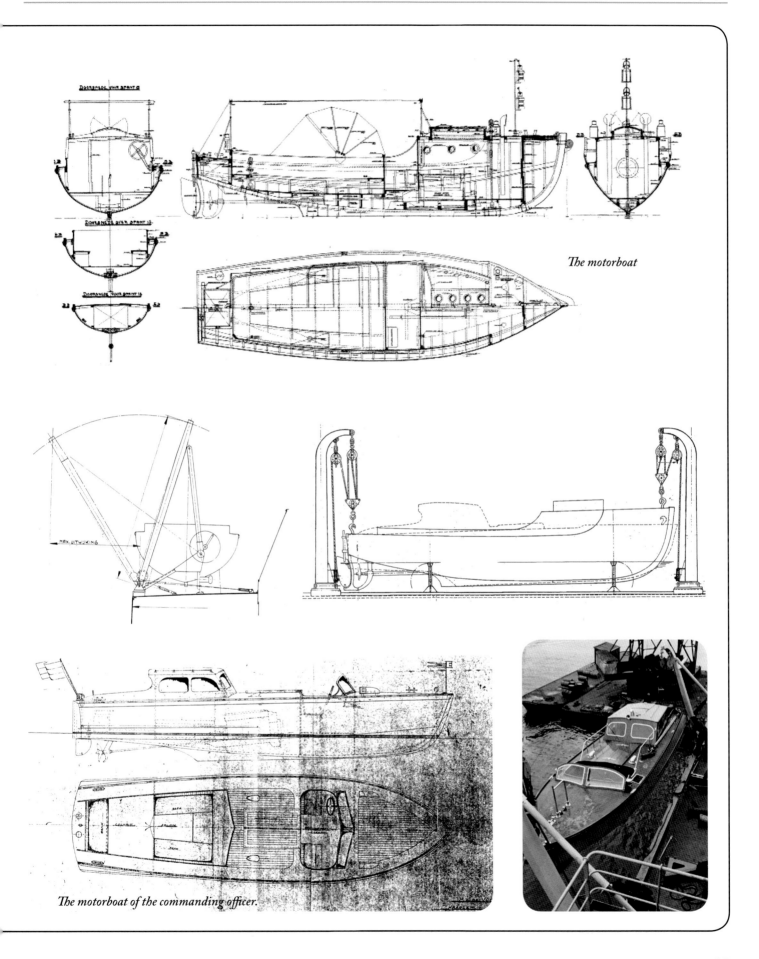

The motorboat

The motorboat of the commanding officer.

Ships name

The name "Tromp" is a remembrance dedicated to two famous admirals who served the Dutch Admiralty from 1607 till 1677:

Maarten Harpertszoon Tromp (April 23, 1598 – August 10, 1653)

and, right, his son, **Cornelis Maartenszoon Tromp** (September 9, 1629 – May 29, 1691)

When the King Leopold III of Belgium visited Amsterdam in November 1938. The naval ships are illuminated in this harbour. (Collection R. Edelenbosch)

Pre-war

The first months HNLMS *Tromp* remained in home waters. The first time the public saw the ship was on 3 September 1938 during the naval review off the coast of Scheveningen as part of the 40 years jubilee of Queen Wilhelmina. On 7 November *Tromp* leaves for Portsmouth for exercises and a short visit. Six days later she left for Naples, carrying out trials in Mediterranean conditions.

The voyage to the Iberian peninsula and the Eastern Mediterranean ends abruptly when *Tromp*, moored alongside during her visit to Lisbon is rammed by the German m.s. *Orinoco* (a 9660 ton passenger ship of the Hapag Line). The ship rams *Tromp* amidships on starboard during which the bow touches the wing of the floatplane. However, the damage to the railing and the hull are reasons to be drydocked for three days for closer inspection. The voyage is cancelled and the ship returns to Den Helder.

In view of the threatening war HNLMS *Tromp* is speedily despatched to the Netherlands East Indies in August 1939. Three weeks later she arrives at Sabang, the first port in the Netherlands Indies for fuelling, and employed under operational control of the CZMNI (Commander-in-Chief Naval Forces in the Netherlands Indies), vice admiral H. Ferwerda.

The ship continues her voyage patrolling along the west coast of Sumatra and 'maintaining neutrality'. When the ship arrives in Surabaya she becomes a unit of the navy organisation of the Netherlands Indies. A sizable part of the Dutch crew leaves the ship and is replaced by Indonesian personnel, mostly conscripts. There after Malay is often spoken, next to Dutch. The crew members signed off are transferred to other navy ships.

Vice admiral Vos arrives, behind him the commanding officer, commander Doorman. (Collection Jt. Mulder)

Air Defence

When HNLMS *Tromp* was commissioned she had four Bofors 40 mm anti-aircraft guns and four Vickers 12.7 mm (0,5 inch) guns. Additionaly her main guns could be used against aircraft.

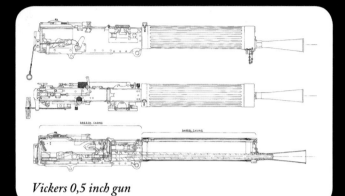

Vickers 0,5 inch gun

The two Vickers 12,7 mm, twin mountings were positioned on the rangefinder-deck.

Pre-War ad of the Bofors Company.

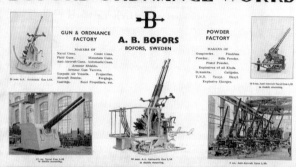

Two Bofors mountings were located aft. The Swedish Bofors 40 mm in the Royal Netherlands Navy was a subject of pre-war technical innovation. It was an excellent mid-range AA-weapon, using the experience of Krupp with semi-automatic guns. A high rate of fire and a relative good "punch". Hazemeyer developed a system for triaxially stabilizing these guns, which meant that among others the impact of the sea state on fire control accuracy was reduced.

The Netherlands Navy had two types in service, the No.3 and No.4. The first design (was mounted on the cruisers *Java* and *De Ruyter*) had a central fire control system, HNLMS *Tromp* had fire control for individual mounts.

The mounting on Tromp.

The original situation.

During an exercise.

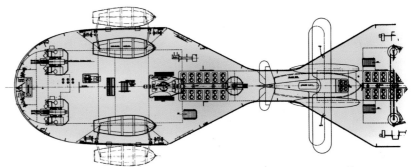

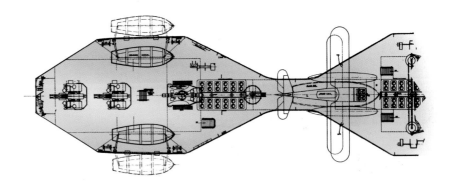

The picture below was taken 1940/41 (note the Fokker plane) in the East Indies. It shows the ship with a reconstructed superstructure aft. This was done to create a greater arc of fire for the 15 cm guns on the quarterdeck. Therefore both 40 mm Bofors have been relocated. While one of them has been placed on a higher platform.

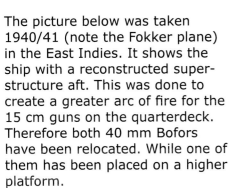

(Photo: Van der Meer Mohr)

World War II

It is 10 May 1940 when just after sundown and while there are guests on board, the message is received that the German army has invaded The Netherlands without any warning. It causes quite a consternation. The surprise attack on The Netherlands makes an enormous impression.

Subsequently the ship is detailed to escort ships in the Indian Ocean and the Pacific due to the presence of German auxiliary cruisers and blockade runners which were active in these waters.

Floatplane

The Fokker C.XI-W was a reconnaissance floatplane designed to operate from warships in the mid 1930s. It was a conventional single-bay biplane with staggered wings of unequal span braced by N-struts. The pilot and observer sat in tandem, open cockpits, and the undercarriage consisted of twin floats. The wings were of wooden construction with plywood and fabric covering, and the fuselage of steel tube, also covered with fabric.

The prototype first flew on 20 July 1935. The aircraft was used to equip the cruisers HNLMS *Tromp* and HNLMS *De Ruyter* while operating in European waters. Although HNLMS *Tromp* had her plane on board when she sailed to the Netherlands East Indies, it was landed after arrival.

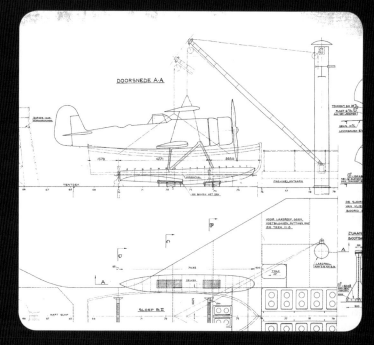

The W11 embarked.

Arriving at Sabang.

In the second half of 1941 HNLMS *Tromp* was temporarily released from the squadron stationed in the Netherlands East Indies and despatched to the south west area of the Pacific. On account of the steadily increasing tension in the Eastern hemisphere after the start of the European war, her presence in that area was deemed desirable. Orders were to protect the Netherlands merchant shipping lanes between the United States and the East Indies. Vital war material was brought to the Indies, while the east-bound ships carried raw materials and other products to feed the rapidly growing armament industries of the United States. This task was interrupted in November 1941 by a fruitless search in the Indian Ocean south of the East Indian Archipelago for the Australian cruiser HMAS *Sydney*, which had disappeared in that area without a trace after having sighted a German armed merchant cruiser. The latter also had vanished completely during or shortly after that encounter. HNLMS *Tromp* was then recalled to the Indies, for it was anticipated that war between Japan and the United States was merely a matter of days, even hours.

Immediately after Japan attacked American and British possessions in South East Asia and in mid-Pacific, the Netherlands Government, exiled in London, made it known through the Governor General of the Netherlands East Indies that it considered this attack was directed at Netherlands colonies also, and that in consequence the Kingdom of the Netherlands was at war with Japan.

From the onset the trade routes to Singapore along the east coast of Sumatra were within easy reach of bombers based on the airfields already prepared or occupied in Vichy French Indo-China, in Siam, as well as on the Malay Peninsula as soon as the enemy had landed his invasion army in that territory. Air attacks and the presence of strong Japanese naval forces concentrated in the South Chinese Sea, threatened to cut the only lines along which reinforcements were able to reach Singapore to support the British forces defending the peninsula. These lines needed immediate protection, but the Royal Navy had not sufficient ships in this area for screening or convoy work. The holding of Singapore and the retention of the Malay island barrier in Allied hands being vital, HNLMS *Tromp* and most of the surface vessels of the Netherlands seagoing fleet in the East Indies were accordingly put at the disposal of the British naval command in Singapore.
This was done also with the submarine divisions which had been ordered to the South Chinese Sea by the C.-in-C. of the Royal Netherlands Navy in the East Indies before the outbreak of the Pacific War.

Japanese forces cut their way with bewildering speed through the British defences on the Malay Peninsula; indeed within a few weeks they were already dangerously near to Singapore, ill-prepared as it was to withstand an attack from the land side. Contemporaneously most of the U.S.-controlled Philippine Islands north of the Nederlands East Indies were falling to the Japanese at the same rate and thus from that direction too the enemy was rapidly approaching the Netherlands colonies.

On the eve of the outbreak of the war the U.S. Asiatic Fleet in that area comprised of only 1 light cruiser, 8 old destroyers and about 27 submarines, with their repair and supply ships.

Fokker C.XI-W

Dimensions:
Length:	10.40 m (34 ft 2 in)
Wingspan:	13.00 m (42 ft 8 in)
Height:	4.50 m (14 ft 10 in)
Wing area:	40.0 m2 (431 ft2)
Empty weight:	1,715 kg (3,781 lb)
Gross weight:	2,545 kg (5,611 lb)
Engine:	1 × Wright R-1820-F52, 578 kW (775 hp)

Performance
Maximum speed:	280 km/h (174 mph)
Range:	730 km (454 miles)
Service ceiling:	6,400 m (21,000 ft)
Rate of climb:	4.8 m/s (940 ft/min)

Armament
1×	fixed, forward-firing 7.9 mm (.31 in) FN-Browning machine gun in forward fuselage
1×	trainable, rearward-firing 7.9 mm (.31 in) FN-Browning machine gun in observer's cockpit

HNLMS Java, seen from the bridgedeck of Tromp.
(Photo NIMH)

Port Blair

Sabang

Western Force

They were withdrawn to Netherlands East Indies and then to Australia as their own naval base Cavite was far too open to air attack. The truth of this was proved on the day war broke out.

Both in valour and fighting efficiency the American sailor was fully the equal of his British and Netherlands contemporaries. During the early months of hostilities aggressive inspiration from the U.S. naval command in this area seemed to be lacking. It would take some time before these forces were given the opportunity of throwing their full weight into the war.

While nearing rapidly the western part of the Netherlands Indies by way of the Malay Peninsula and the Southern China Sea, a second Japanese advance emanating from the Mandate-islands and the partly captured Philippines developed, menacing the whole eastern part of that archipelago. These two advances, of which the second was divided into two thrusts, repetively to the east and west of the island of Celebes were in fact a pincer movement to crush the weak Allied forces defending the Malay barrier.

In their move southwards the well-co-ordinated Japanese naval, air and amphibious forces employed the army airfields on the islands they captured. To be used as stepping stones for their bombers and fighter aircraft. By this means adequate air support was available to the enemy's next naval and amphibious operations.

So the East Indies -last Allied rampart in South East Asia- were invaded from three sides. The hurriedly set up Allied combined command headed by the British general Wavell did its best to counter the offensive, but as it was ill prepared for this heavy task valuable time was lost before this new organization took effect.
After some deliberations the point of view was accepted of the Netherlands naval

C.-in-C., vice admiral Helfrich, who had consistently advocated leaving convoy and patrol duties to lighter naval units and employing the larger ships, like cruisers, in a more aggressive role. After two months of war, however, it was far too late to obtain proper results with a hastily improvised Allied squadron as there was no time left in wich to train British, American and Netherlands ships overnight united as a co-ordinated, manageable fighting force. Neither a joint communication system nor a common naval signal code yet existed. As a result, only the basic tactical manoeuvres could be carried out without the possibility of misunderstanding which could easily end in a disaster. Moreover the Allied forces had already incurred such heavy losses that the skies above the archipelago and the surrounding seas were dominated by Japanese bombers with their fighter escorts without any likelihood of effective opposition.

HNLMS De Ruyter

HNLMS Java

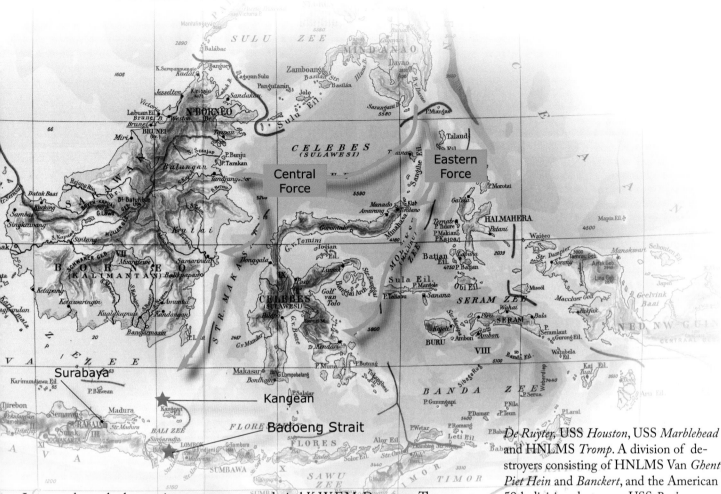

It was under such gloomy circumstances that the Allied striking force was established on 3rd February, and the next day it was despatched on its first mission. The force commander, rear-admiral Karel Doorman, Royal Netherlands Navy, himself an experienced airman, (the brother of the first commanding officer of *Tromp*) had continually urged that his force be supported with sufficient air cover from the nearest airfield but no such assistance could now be given because long range fighters were no longer available.

On 3rd February 1942 *Tromp* is incorporated into the Combined Striking Force (CSF) under the command of rear-admiral K.W.F.M. Doorman. The squadron assembles near the Gili's south of Madura. During the night different formations of this CSF sail separately to the rendez-vous where the fleet has to assemble at 05:00 hours. The Allies are of the opinion that the Japanese are setting course for the Java Sea from Balikpapan (East Borneo), with the ultimate aim to invade East Java, or an attack on the southwesterly part of Celebes. A scouting action must be made to confirm.

When Japanese ships are reported in the southern part of Macassar Strait, the Combined Striking Force sets sail. The Squadron consists of the cruisers HNLMS *De Ruyter*, USS *Houston*, USS *Marblehead* and HNLMS *Tromp*. A division of destroyers consisting of HNLMS Van *Ghent*, *Piet Hein* and *Banckert*, and the American 58th division destroyers USS *Barker*, *Bulmer*, *Edwards* and *Stewart*.

Near Kangean, the Combined Striking Force engages Japanese air force planes in which the USS *Houston* and *Marblehead* are severely damaged. The Dutch ships suffer only slight damage.

When the squadron, in the dark of the night of 15 February 1942, enter Stolze Strait there is another heavy rain shower sweeping surface like a veil. The fleet now consists of five cruisers: *De Ruyter*, *Java*, *Tromp*, *Exeter* and *Hobart*. Ten destroyers: *Banckert*, *Piet Hein*, *Kortenaer*, *Van Ghent*, *Barker*, *Bulmer*, *Pillsbury*, *John D.Edwards*, *Parrott* and *Stewart*.

HNLMS Tromp

HNLMS Banckert, Piet Hein, Kortenaer and Van Ghent

Crew of gun III posing in 1941.
(Photo Van der Meer Mohr)

In the bad weather the destroyer HNLMS *Van Ghent* runs on the Bamidjo reef and severe flooding started. When a fire breaks out as well the crew had to give up. The *Banckert* transferred personnel and stores.

The Combined Striking Force courses westward and is spotted by a Japanese 'plane from the cruiser *Chokai*. Doorman reacts immediately and asks repeatedly for air cover but in vain.

Just before 12:00 hours the first Japanese bombers start their attack. By fast manoeuvring and anti aircraft fire the attacks are beaten off. During this encounter *Tromp* fires her anti aircraft guns once. There are no hits, but a number of bombs fell frighteningly close. Elsewhere in the archipelago, the invasion of south Sumatra takes place which ultimately would lead to

the surrender of Singapore. All available aircraft had been deployed for defence.

This day would enter history as Black Sunday. The lack of air cover for the Allied squadron is again emphasized by this. Now that the element of surprise is lost, the intended raid on the north coast of Banka is cut short. The fleet takes the opposite course to Stolze Strait. The air attacks continue however. USS *Barker* and *Bulmer* are damaged by 'near misses' and leave the squadron.

On 19 February 1942 a formation of Japanese warships and troop transports anchored in Strait Badoeng, the waterway between the Southern point of Bali and the island of Penida.

--*Later it would appear that the scouting reports were not correct. During the night there were only three Japanese destroyers in the narrow straits* --

At the time the decision for this operation was taken, the Allied squadron had only just returned from its fruitless venture in the Southern China Sea, and its ships were refuelling in different ports, so only part of the force could be assigned to this task.

Orders were given that the Netherlands cruisers *De Ruyter* and *Java*, a short while after followed by three destroyers, were to enter the Straits after dark from the south, with the intention of surprising the enemy and while proceeding at full speed along the line of anchored ships to inflict heavy damage. After this *Tromp*, preceded by another four destroyers, was

USS Houston

USS Barker, Bulmer, Pillsbury, John D.Edwards, Parrott and Stewart

to approach the positions just left by the first force in order to finish off what remained of the enemy who was assumed to be in confusion. The second group also was to come from the south as might be expected that the enemy was only looking in the direction in which the attackers had vanished into the darkness. Finally, two groups of motor torpedo boats from Surabaya were to arrive for the kill.

During the night of 19 to 20 February 1942 the attack commenced. The Combined Striking Force was divided into two formations, of which the second group would attack three hours after the first. The first group did indeed surprise the enemy. HNLMS *Java* fires three star-shells in different directions and noticed confusion amongst the crew on one of the destroyers. When the cruiser sails past she fires her guns. The enemy sustains a great number of hits but does not sink. The other ships of this first group also participate. It all lasts a few minutes only, after which the first group of the Combined Striking Force disappears into the darkness.

When the second group with HNLMS *Tromp* and the 58th Division destroyers commence attack there is no question of a

Forecastle seen from the foremast.
(Photo Van der Meer Mohr)

surprise effect. Soon afterwards the cruiser is caught in a searchlight over starboard and take serious punishment. In the engagement the ship receives 11 hits (10 killed and 30 wounded), from the Japanese

destroyers *Oshio* and *Asashio*. One of the Japanese destroyers gets a direct hit on her bridge. In the mean time the destroyers *'Arashio'* and *'Michishio'* enter Strait Badoeng.

A view aft, shortly before WW2.
(Photo NIMH)

A peaceful photo near the Sydney Harbour Bridge. (Photo NIMH)

Colombo

Since the surrender of the Italian Navy the British Eastern Fleet had been considerably strengthened and made repeatedly offensive sweeps in the eastern part of the Indian Ocean.

On 4 January 1944 HNLMS *Tromp* leaves the South West Pacific Area after which she arrives in Colombo on 13 January to be added to the Eastern fleet. Trincoma-lee, an important naval base of the Allies, becomes the homeport of *Tromp*.

On 16 April 1944 the light cruiser sails with the Eastern Fleet for the first time for an operation in Indonesian waters. The aim is an attack on the harbour of Sabang and the airfield of Llonga.

Sabang is situated at a point where all trade routes come together, those from the Indian Ocean and those from the Malacca Strait. Long before the war the Dutch had installed, at the Northern side of the bay, extensive provisions to store and repair ships. Although these had been damaged when the Japanese landed there in 1942, but they also recognised the importance of the position and repairing started rather quickly again. Much had been done to repair the harbour and the airfield. In addition a good defence had been set up against atacks from the sea or from the air. The fleet consists of twentyseven warships, from six different countries, divided into two groups: 'Task Force 69' and "Task Force 70'. The 'planes from the carriers in these TF's had an important task in this operation.

19 April 1944 – Operation Cockpit (The attack on Sabang)

In the early morning thirtyeight Barracuda, Avenger and Dauntless bombers, accompanied by fortyseven Corsair and Hellcat fighters from the aircraft carriers carried out an air attack on Sabang and the nearby airport.

The action is a great success. Two Japanese fighters hit, an escort ship is set on fire, two merchant ships of more than 4000 tons each are hit and an oil storage tank catches fire. Several buildings are hit among which are the shipyard, barracks,

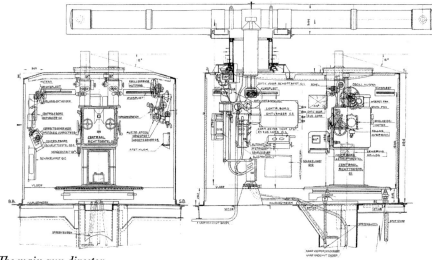

The main gun director.

For the crew of *Tromp* it is a tense moment when for the first time after two years the mountains of Java rise above the horizon. The island where many of them had to leave their families behind.

22 July 1944 – Operation Crimson (The second attack on Sabang)

The British Eastern Fleet leaves Trincomalee in order to carry out a bombardment from the sea on Sabang so that the harbour is made unserviceable for the Japanese fleet. The hope is to regain the allied mastery of the Indian Ocean. It

is the prerequisite before actions by sea can be undertaken against Burma and Malacca. *Tromp* is assigned with three RN Q-class destroyers to enter the bay of Sabang to carry out close range bombardments. Targets are ships in harbour, oil tanks and the workshops of the Sabang Company. While the ships are in the bay bombardment by the Allied ships will be postponed.

After the squadron has opened fire, speed was increased. HMS *Quilliam*, HNLMS *Tromp*, HMS *Quickmatch* and HMS

In Fremantle Harbour, 1943.

power stations. In many places started. On the airport hangars and other buildings are hit and twentytwo planes are destroyed on the ground. The Japanese started a counter attack by sending in three torpedo planes, but these planes are downed by the fighters from USS *Saratoga*. Only one Hellcat of the Allies is downed. The pilot is picked up from the sea by the submarine HMS *Tactician*. After the action the return voyage went off without incidents worth mentioning.

17 May 1944 – Operation Transom (The attack on Surabaya)

The attack on Surabaya started with eightyfive aeroplanes (fortyfive Avenger and Dauntless bombers, accompanied by forty Corsair fighters) The targets of the aeroplanes from USS *Saratoga* and HMS *Illustrious* are the harbour and oil refineries of Surabaya.

With a new US grey coat.

The colour scheme till 1944

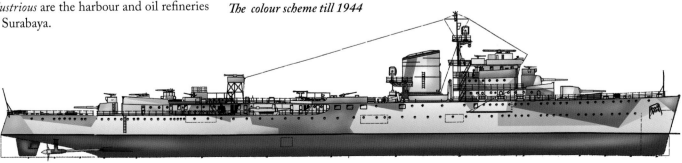

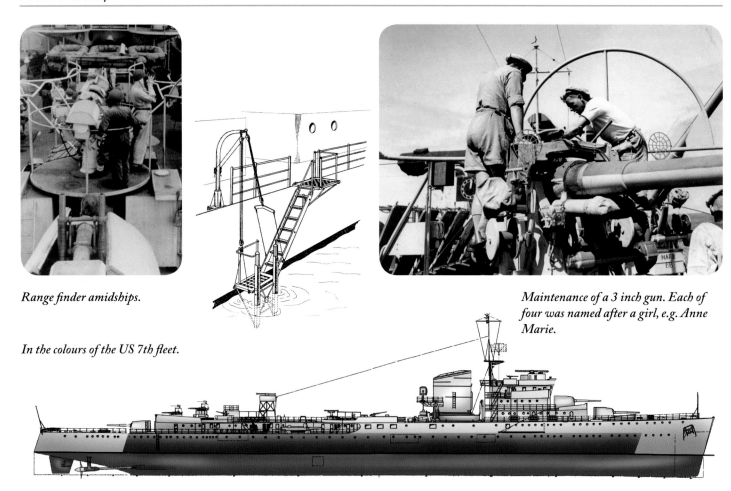

Range finder amidships.

In the colours of the US 7th fleet.

Maintenance of a 3 inch gun. Each of four was named after a girl, e.g. Anne Marie.

Another modification, the searchlight platform at the funnel has been lowered and is now used for machineguns. (Collection R. Edelenbos)

View from the platform of the fore mast.

Replenishment at sea.
(Photo NIMH)

Below:
Fremantle. (Photo NIMH)

Quality steamed at 22 knots into the bay of Sabang, in that sequence. Of these ships Tromp carried the heaviest armament and therefore had the freedom to fire at the most important visible targets. The RN destroyer HMS *Quilliam* under command of captain R.G.Onslow (DSO with three bars), leading formation to facilitate navigation.

06:45 hours. An ancient tradition, the national ensign is hoisted in top, HNLMS *Tromp* is ready for battle. When the harbour comes in sight the direction finder is used to locate the targets. Meanwhile guns I and II follow the direction finder and on the aftship gun III is trained in the

most forward position on starboard side. By air reconnaissance earlier most of the anti aircraft gun sites had been located. On the photographs almost no coastal batteries had been disclosed as they had been camouflaged by the thickly foliaged surroundings of hundreds of palm trees. For that reason the distances were measured on the roofs of some warehouses.

06:57 hours. *Tromp* opens fire with the 15 cm guns. Each mounting fires with one barrel. After two minutes salvoes (double shots) are fired by the foreward mountings.

In the mean time the coastal defence artillery is being fired upon by the battleships, which are stationed outside the bay. The result is that friendly hits are difficult to observe. Only when the enemy fires the gunnery control can spot a position by the flash of their guns. At first little resistance is experienced.

After an alteration of course to starboard, at **07.05 hours** gun III also joins battle. Several targets are hit. In front of a shipyard an ammunition ship alongside is hit, it explodes to fragments. Immediately guns are trained at the coal sheds and workshops. Distance to shore has grown close to enable gunfire support of the 7.5 cm guns and later the 40 mm's. Gunfire from the western shore of the bay is answered by the 40 mm guns. When the ships negotiate a small cape, a Japanese gunfire, initially badly trained, hits one of the British destroyers. The crew of *Tromp* observes one hit after another on the British ship, one of which hits the

HMS Quilliam, Quickmatch and Quality

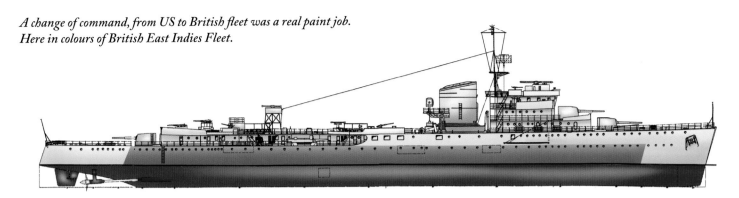

*A change of command, from US to British fleet was a real paint job.
Here in colours of British East Indies Fleet.*

bridge. HNLMS *Tromp* shifted her line of fire and silences the Japanese battery.

Tromp nears the shore to approximately 2000 metres and received a first hit at 07:13. An aerial is destroyed and a shell hits the fore range finder. For one of the veterans (Badoeng Strait) reason to remark: *"And there goes another one"*. All guns are firing now. A pandemonium of hell broke loose.

Only after the battle it will be assessed *Tromp* has been hit many times. Luckily nobody is wounded and none of the projectiles has exploded. A 12 cm shell had entered the starboard whale corridor and without exploding lands on armour grating above the boilers. This is obviously very dangerous because the extreme heat could lead to an explosion. The shell rolls back and forth slowly. C.P.O.-engineer P. Steenaard wraps himself in wet bags and in spite of the great heat enters the funnel gallery. He wraps the 30 kilo heavy shell in a wet bag and throws it quickly overboard. Another shell crashes past the feet of the executive officer (XO), lieutenant K. van Dongen and hits the aft windlass. This shell does not explode either.

Less fortunate are the other ships: HMS *Quilliam* is hit aft killing one and wounding four. HMS *Quality* receives a hit which wounds a number of crew killing one.

At **07:20 hours**, when the 'Inshore Force' leaves the bay a smoke screen is laid and speed increased to 28 knots. *Tromp* has fired 205 rounds 15 cm high-explosive shells and 51 rounds with the 7.5 cm guns. Both 40 mm mounts have fired 770 shells.

–During this action the ship was in the bay for 25 minutes during which 4 hits were received. On the way back, during surveying the damage suffered a Japanese shell of 13 cm was found in a colander in the galley. This one had not exploded and was thrown overboard by quartermaster S.T. Hendriks (who was killed later off Port Blair). On the aft ship a splinter of a 3 inch shell is found. This was used as a paperweight for a long time. (Now part of the collection of the Marinemuseum at Den Helder.)–

First rounds being fired during Operation Crimsom. (Photo NIMH)

The story goes that on the way back *Tromp* received a signal from the admiral: *"Well done"* which is followed by a general signal from the flagship with the text: *"We got them with their kimonos up"*, a variant of the well known saying: *"to catch a person with his pants down"*.
Due to this successful action many congratulatory messages are received. As thanks for the support given, the commanding officer of HNLMS *Tromp* receives from captain R.G. Onslow, commanding the British destroyers, a

silver salver with inscription. The ship was also visited by rear admiral A. Read (Commander 4th Cruiser Squadron).

Her Majesty the Queen accorded captain F. Stam and C.P.O.-engineer P. Steenaard the Bronze Lion and nine officers, petty officers and crew the Bronze Cross.
It was thought amusing that a few days later the Japanese radio in Tokyo broadcasted an English language eye-witness report about the destruction and sinking of the ships that had attacked the bay of

Sabang. It was the third time that Radio Tokyo 'sank' *Tromp*.

The months November and December of 1944 HNLMS *Tromp* spends in the yard in Sydney. Apart from maintenance, the British radar is substituted by two sets of American manufacture, a radar of the type SG for surface warning and a radar of the type SC for air warning. In addition the ship receives a British Type 285 gunnery radar for the main armament and two Type 282 radars for the 40 mm mounts.

(Photo NIMH)

27 April 1945 – Operation Dracula (landing near Rangoon)

Six convoys sail in order to land a British force with the intention to capture Rangoon. Part of this action was Operation Bishop, the covering of the landing operations and at the same time, the bombing of the Nicobar Islands to prevent possible sorties by the Japanese from the Andaman and Nicobar Islands. The commanding officer of *Tromp* is Senior Officer of the group and thus was in command. While planes are bombing the airfield of Car Nicobar, the ships fire at various targets after which course was set to the Andamans. During the bombardement of Port Blair a 15 cm shell explodes on leaving the muzzle. Splinters wounded fourteen crew and also collateral damage has been caused, even the port side torpedo tubes, amidships, are slightly damaged.

After Rangoon has been captured on 3 May, a third shelling of Ross Island follows. Two days later it was 'Sunday Routine' on board and the ship was dressed in celebration of the end of the war in Europe. Later the ship takes part in a 'Shipping sweep' in the Andaman Sea. On 16th May 1945 deployed in Malacca Strait, when the Japanese cruiser *Haguro* is sunk by Allied destroyers.

With HNLMS Jacob van Heemskerck during exercises. (Photo Pennewaard)

Torpedotubes

The light cruiser was equipped with 6 torpedotubes for 21 inch (53 cm) torpedo's in triple mountings on the maindeck. Also she carried 6 spare torpedo's.
When *Tromp* sailed to the East Indies, she carried 12 brandnew V 53 torpedo's. In spring 1945 the torpedotubes were, in Sydney, replaced by British tubes for launching the Mk 9 torpedo's.
The tubes were unshipped in 1946.

V 53 torpedo

Supplied:	Whitehead, Weymouth
Length:	7,2 meter
Diameter:	21 inch
Weight:	1,650 Kilo
Warhead:	300-350 Kg TNT
Range:	4000 m at 45 kts
	12000 m at 28 kts

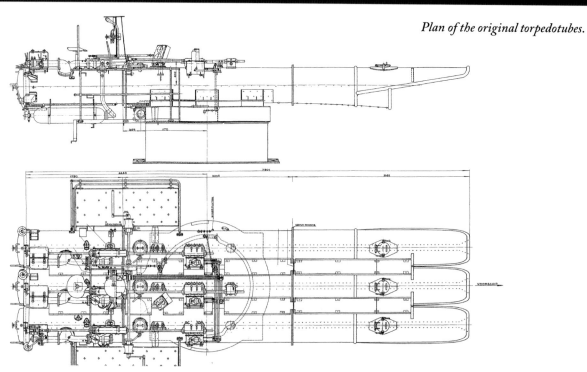

Plan of the original torpedotubes.

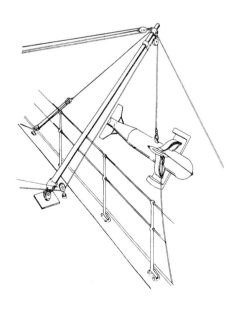

In May it was decided *Tromp* should operate with the US 7th Fleet (admiral Thomas C. Kinkaid). In June the cruiser joined Task Group 74.2 consisting of the cruisers USS *Montpelier* and *Denver* and four Fletcher class destroyers.

Tromp was busy in preparing and carry out **Operation Oboe II (the landing near Balikpapan)**.

During this operation *Tromp* is able to destroy two shore batteries. In the entrance to Balikpapan gun support is given to protect mine disposal operations and to underwater demolition teams. HNLMS *Tromp* postponed operations after all ammunition has been used.

End of war

The commanding officer receives a message that his ship has been detached from the US 7th Fleet and he has to report to the Commander-in-Chief East Indies Fleet for further orders. On 15th August the Japanese emperor granted permission to surrender. A month later, on 16 September, HNLMS *Tromp* arrives on Tandjong Priok roads. The cruiser was the first Netherlands warship that returned to the Netherlands Indies after the war.

The capitulation of Billiton is signed on board *Tromp* on 12 October. The Japanese 1st Lieutenant Yamaguchi Yashiro unconditionally surrenders the island to the C.O. of *Tromp* at Tandjong Pandan.

Tromp in the last stage of the war.
(collection Jt. Mulder)

In the mast the ship wears her pennantnumber D28 (RN code flags).
(Collection Jt. Mulder)

Tromp in the final days of her commission, with a new post-war grey coat and modified funnel.

Returned to The Netherlands

The ship berthed in Amsterdam in the morning of 3 May 1946. The end of a deployment of nearly seven years. And, two days short, a year after the liberation of The Netherlands.

Shouts of recognition, cries of welcome, tears and cheers and many expressions of joy. After seven years she has become the renowned *Tromp* and has earned a great reputation. A ship which continued to fight for freedom.

A heading in the magazine "De Spiegel" (The Mirror) announced: *"The pride of our Netherlands Navy has returned after seven years of absence"*.

After a short stay in Amsterdam, the ship sails to Rotterdam for maintenance and overhaul after which, on 1 July 1948 she was ready for service. The funnel has been altered and now has a wide horizontal edge with two curved vertical plates fitted on. It must decrease the precipitation of soot on the aft deck. It appears to be an improvement. The boiler room crew is much less happy. It appears that sometimes this modification can be life threatening. In stormy weather draught develops down the funnel. If, at such a moment, the peep hole of a burner is opened for inspection, the hot burning gases blow out. Sometimes combined with a jet of flame. Many a crew member became a little more bald due to this. However, no further adaptations were made, only the use of safety glasses was made compulsory when inspecting the burners.

From July 1948 the armament of the ship was:
6 guns of 15 cm in twin mountings
4 guns of 7.6 cm
8 machine guns 40 mm
4 oerlikons of 20 mm
6 (2 x 3) torpedo-tubes 53.3 cm

The start of the second week of the new year becomes a memorable day. On that day the Royal Mention by Order of the Day is awarded by Royal Decree of 8 January 1949, number 36.

End of June 1949 participating in fleet exercises of the Western Union in the Channel and the Bay of Biscay. Prelude to NATO.

At the end of her career she is employed as training ship for the Technical Training Royal Netherlands Navy (TOKM) in Amsterdam.

Ship's badge

Since 1940 the Royal Neth. Navy has enjoyed a heraldic tradition in the form of its badges. Often created by the crew, sometimes more or less official. This changed in 1950 when the heraldry of the navy has been described in regulations.

The badge of HNLMS *Tromp* was allocated in August 1952 and bears the coat of arms of the admirals Tromp (see page 24). Father and son, who both won great distinction commanding the Dutch fleet.

Between 1950 and 1955 the pennant-number assigned was C 804.
(Photo NIMH)

Translation	Graphic design
Frits Kumer	Jantinus Mulder
Author	**Publisher**
Jantinus Mulder	Lanasta
Corrections	
Henk Visser	

First print, August 2012
ISBN 978-90-8616-191-1
NUR 465

Contact Warship:
Slenerbrink 206, 7812 HJ Emmen
The Netherlands
Tel. 0031 (0)591 618 747
info@lanasta.eu

Lanasta

Tromp in 1948.
(Photo NIMH)

References
- Huibert V. Quispel: *The job and the tools,*
The Netherlands United Shipbuilding
Bureaux Ltd 1935-1960.
- G.H. Kleinhout, J.C.I. Landegent, Jt.
Mulder & C. Sybesma: *De Tromp en haar*
Trompers, Lanasta 2003.